THE
GLASS ARK

The glass arches spanning the farm area of Biosphere 2 catch the moon's glow as it rises over the Santa Catalina Mountains at twilight. Biosphere 2 brings a new element of beauty to the Sonoran Desert as well as a glimpse into the future.

THE
GLASS ARK

THE STORY OF BIOSPHERE 2

BY

LINNEA GENTRY & KAREN LIPTAK

VIKING

VIKING

Published by the Penguin Group
Viking Penguin, a division of Penguin Books USA Inc., 375 Hudson Street,
New York, New York 10014, U.S.A.
Penguin Books Ltd, 27 Wrights Line, London W8 5TE, England
Penguin Books Australia Ltd, Ringwood, Victoria, Australia
Penguin Books Canada Ltd, 10 Alcorn Avenue, Toronto, Ontario, Canada M4V 3B2
Penguin Books (N.Z.) Ltd, 182-190 Wairau Road, Auckland 10, New Zealand
Penguin Books Ltd, Registered Offices: Harmondsworth, Middlesex, England

Published in simultaneous hardcover and paperback editions by Viking Penguin,
a division of Penguin Books USA Inc. 1991

10 9 8 7 6 5 4 3 2 1

A ProMundo & Company Production

Library of Congress Cataloging-in-Publication Data
Gentry, Linnea
    The glass ark / by Linnea Gentry and Karen Liptak.
        p. cm.
    Includes bibliographical references and index.
    Summary: Describes the Biosphere 2 project which has created a closed
environment intended to duplicate life on Earth in a way that would facilitate
future space colonies.
    ISBN 0-670-84173-0
    1. Space colonies—Juvenile literature. 2. Closed ecological systems (Space
environment)—Juvenile literature. 3. Biosphere 2 (Project)—Juvenile literature.
[1. Biosphere 2 (Project) 2. Closed ecological systems (Space environment)
3. Space colonies.] I. Liptak, Karen. II. Title
TL795.7.G48   1991
918.8'04—dc20   81-25328
                                                                         CIP
                                                                          AC

Printed in U.S.A.

Cover photographs: diver by Gonzalo Arcila; Biosphere II model, children, and
spaceframe by C. Allan Morgan; Astronaut by NASA, courtesy of Space Imagery
Center— University of Arizona.

Contents

Introduction 7

Chapter 1. A Modern-Day Ark 9

Chapter 2. Building the Glass Ark 21

Chapter 3. All Aboard 37

Chapter 4. The Place for People 57

Chapter 5. The New Biospherians 69

Chapter 6. Visions of Tomorrow 81

The Hard Facts 87

For Further Reading 88

Glossary 89

Index 92

Introduction

There is an ancient story told in the Bible about a great flood that covered the world long ago. All of the people and animals were threatened with destruction. But there was one good man named Noah whose family God wanted to save. So He warned Noah that the great flood was coming and told him to build a huge ark.

Noah and his family built the ark and loaded it with food. They also took on board a male and a female of every kind of animal—eagles, sparrows, snakes, sheep, giraffes, mice, leopards, platypuses. As it began to rain, everything came up the gangplank, two by two. It must have been a wonderful sight!

The rain poured down for forty days and forty nights until the world was entirely under water. Although the sun finally came out, it would be many months before the waters receded. Meanwhile, the ark with all its inhabitants floated alone over the deep.

One day Noah sent out a dove, but she found no place to land. Later he sent her out again. This time she returned with an olive leaf in her beak. When Noah sent the dove out a third time, she didn't return. He knew that she had found solid ground and that soon the water would dry up. And indeed, after almost a year in the ark, Noah and his family and all the animals were able to walk out onto dry land and begin a new life on Earth.

SCOTT McMULLEN

Seen from the north side, Biosphere 2's rainforest basks in the sunlight of early spring. Built for a hundred-year lifespan, this giant airtight greenhouse shelters an exciting experiment that may change the way we think about our home planet and our explorations in space.

Left: These young visitors examine some of the seashells on Biosphere 2's beach before the structure is closed.

C. ALLAN MORGAN

CHAPTER 1

A Modern-Day Ark

The story of Noah's ark is a wonderful legend. There are many similar legends told all over the world about the Great Flood and the saving of the people and animals. These legends remind us that Earth's inhabitants are precious and fragile, just like our planet itself, and are worth being saved and protected.

Now Earth is in danger once again, but not from a Great Flood. Overpopulation and industrial growth have brought pollution, global warming, and acid rain which in turn have caused changes in our atmosphere, our oceans, our plant and animal life, and our food supply. All of these problems are caused by people. But they can also be solved by people. Scientists, engineers, environmentalists, and other concerned people all over the world are studying Earth to learn how to save it.

One of the most unusual and daring experiments to learn more about our planet and its problems is taking place in the desert of the American Southwest. Located near the mountains not far from Tucson, Arizona, it is one of the most amazing structures ever built by human hands.

It's not just the location or the shape that is so unusual—it's actually the largest glass greenhouse ever made. It covers an area equal to three football fields and reaches a height of eighty-five feet. It's made of glass and steel and is sealed so tightly that no air or moisture can go in or out. That in itself is a major accomplishment.

But most incredible of all, this sealed greenhouse is the home of a very special group of

NASA

In this photograph of Earth, taken from space during one of the missions to the moon, the continents of Africa and Antarctica appear through swirling clouds. Some people have suggested that a better name for planet Earth would be "planet Ocean" because water covers so much of its surface—over 70 percent, in fact.

inhabitants. A crew of four men and four women from three countries live inside its glass walls with four goats, three pigs, numerous chickens, and 3,800 other species of animals and plants gathered from around the world. They are part of an amazing experiment that is trying to imitate life on Earth. Sealed inside the glass is a miniature world complete with a tropical rainforest, a savannah (a grassy plain), a marsh, a desert, a farm, a village, and even a miniature ocean.

In the fall of 1991, the doors were sealed shut for the first long-term closure. For two years no supplies will come in from the outside world, just sunlight and electricity. Everything else that they need to survive is inside with them. Food, water—everything! Its creators call this new world "Biosphere 2."

A biosphere is a system of living things successfully working together to keep their envi-

Here the seven areas, or biomes, inside Biosphere 2 are shown clearly. On the left are the farm and the Human Habitat. On the right are the wilderness areas— rainforest, savannah, desert, marsh, and ocean. The large white domes enclose Biosphere 2's "lungs," huge balloon- like sacks that inflate and deflate to equalize air pressure inside the glass.

ILLUSTRATION BY ELIZABETH DAWSON

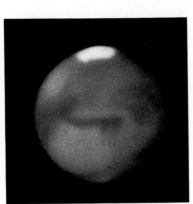

Earth is the only planet in our solar system which supports a biosphere. All the other planets are either too far away from the sun and too cold, like Mars (*right*), or too close to the sun and too hot, like Venus (*left*). Someday we may have the technology to build biospheres on other planets. Even today people are studying the possibility of a biosphere colony on Mars.

ronment going. Biosphere 1 is a name some scientists use for Earth—the only natural biosphere we know of. Until this experiment began, no one had ever tried to create a complex biosphere before—not even Noah!

Biosphere 2 is actually a miniature copy of our home planet, Earth. The plants and animals, the soil and water all play a part in maintaining the delicate balance of life inside, just like on Earth.

Of all the planets in our solar system, Earth alone is just the right distance from the sun to support life. And there's life almost everywhere on Earth. Biosphere 1 reaches from the top of the atmosphere to the bottom of the ocean and deep underground. It includes every plant and animal—all the living things we can see, as well as the ones we can't, such as the bacteria and the tiny organisms called microbes. It includes all the water and soil, as well as the wind and rain. It is everything that is alive or necessary for life.

Everything in Biosphere 1 interacts in a complex web of life. All day and all night, living organisms are working together so that everyone may live. Yet they don't even realize that they help each other. For example, right now ants, termites, worms, and other creatures are

Biospherian candidate Stephen Storm examines some early experiments in growing plants from single plant cells at the Biosphere 2 Tissue Culture Laboratory. Someday space explorers may be able to carry a rich selection of Earth's plantlife into space with them using these techniques.

busy breaking apart dead plants and animals for their own uses. By cleaning up old and dead matter, they also help make room for continuous new life and recycle precious nutrients.

Scientists admit there are still many mysteries about how life on Earth stays in balance and survives. That's one reason why Biosphere 2 is so important. It's a tool to help us solve those mysteries and discover how we can take better care of our planet.

Its creators did not want just a functioning copy of Earth in a test tube, however. They wanted an active, well-balanced community of happy, healthy inhabitants. So Biosphere 2 was designed as a separate world where plants, animals, humans, and microbes will not only survive, but will actually thrive!

Scientists are able to watch the animals and the plants inside the closed world of Biosphere 2 very closely. They can study how the plants and animals adapt to change and interact with other species. They can also test new ideas for cleaning our air and water.

Biosphere 2 also plays an exciting and important role as a model for future space stations. Its planners designed it to mimic some of the conditions that humans will face on alien worlds. The life-support systems are part of this. If we ever expect to live on other planets or in space colonies, we must build life-support systems to help us survive. These are *very* important to the health of all living creatures, whether palm trees or butterflies or humans. And they must nourish good mental health as well as good physical health.

People have been learning how to make life-support systems for a long time. Not only do astronauts in space have to take enough food with them, their spaceships must also carry an imitation atmosphere. Scuba divers must do the same when they go underwater. They carry miniature atmospheres in their oxygen tanks!

Answers to questions about ecology and life in space are just some of the fascinating things to learn from this experiment. Someday we may be able to use biospheres as scientific research stations in the depths of the ocean or in Antarctica. They may also be used as recreation centers for people in cities or as special homes for rare and endangered species. Schools may some day use these mini-Earths as living classrooms in which to learn about our planet. Perhaps you will think of ways to use them that scientists don't even suspect now! There are many possibilities for Biosphere 2 and future biospheres. To fulfill these possibilities, the scientists, engineers, and designers of Biosphere 2 had to solve tremendous problems before the experiment could even begin. But before we can understand the enormous undertaking of building a "glass ark," we must first understand how Biosphere 1 works.

There are about 10,000 species of ants on Earth. These important insects live in colonies all over the world, except for extremely cold areas and underwater. The planners of Biosphere 2 had to choose species of ants that would process soil and dead organic matter without harming other Glass Ark inhabitants.

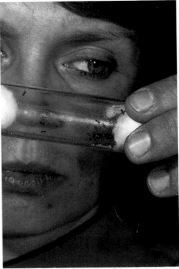

C. ALLAN MORGAN

13

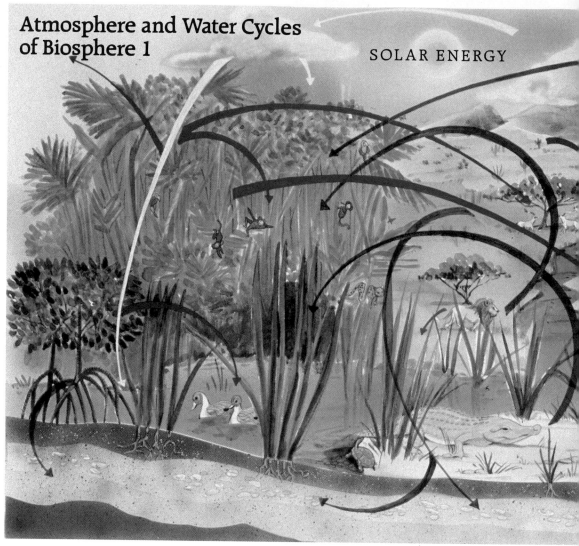

SOLAR ENERGY

ILLUSTRATION BY JANE BARTON

The continuous cycles of water, carbon dioxide, oxygen, ammonia, and nitrogen moving through Earth's sun-warmed atmosphere and its organic material, or biomass, are crucial to the success of life as we know it.

Life on our planet is made possible by many different but interconnected activities. This chart shows how all the different activities on Earth work to support life.

The sun is our planet's power supply. Through the process of photosynthesis, green plants combine sunlight with water and carbon dioxide (one of Earth's gases) to make their food. As a waste product, plants release another very important gas, oxygen. Animals breathe the oxygen made by the plants. During the breathing process, animals give off carbon dioxide. Plants breathe this in and give off more oxygen. It's a continuous cycle. All over

the world, the animals breathe in what the plants breathe out and the plants breathe in what the animals breathe out.

This cycle of oxygen and carbon dioxide is part of the Earth's larger air cycle. Earth's atmosphere is always moving. Some of the movement is caused by Earth's rotation. Some is caused by temperature changes. This temperature-controlled movement, called convection, is the natural process in which air is heated by the sun and then rises.

At the same time, water is constantly being recycled. Water in the atmosphere produces rain. The rain joins rivers which feed into the

Many of Dr. Clair Folsome's early "ecospheres," on display in a Biosphere 2 laboratory, are still alive, some of which have been sealed for over twenty years. After experimenting with seawater and algae, Dr. Folsome added tiny shrimp to his ecospheres. Although the shrimp do not reproduce in such a tiny world, they have lived for several years sealed inside their flasks. Tiny glass worlds like these eventually led to the development of Biosphere 2.

lakes and oceans. From the lakes and oceans, the water evaporates into the atmosphere once again. Dead plants and animals are also constantly being recycled. The complex recycling of plants and animals is called the food web. The chart on the right helps explain the process in pictures.

Humans have been studying these cycles for many centuries to understand how they work. More recently, we have begun to re-create them. Back in 1968, a scientist at the University of Hawaii, Dr. Clair Folsome, scooped up some seawater into a bottle. The seawater contained algae (AL-gee) and tiny microbes. He sealed the bottle so tightly that no air could get in or out and then set it on the windowsill in his laboratory. To his surprise, the algae and the microbes didn't die. They continued to live

KENT WOOD

The Food Web in Biosphere 1

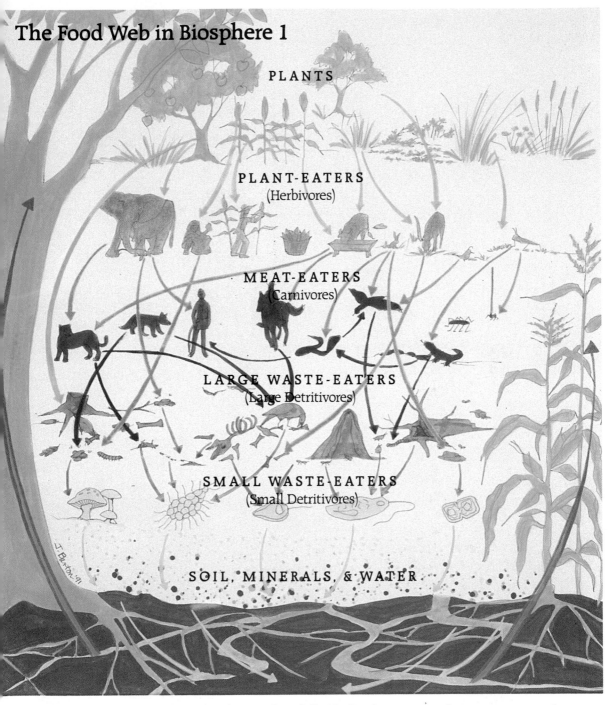

PLANTS

PLANT-EATERS
(Herbivores)

MEAT-EATERS
(Carnivores)

LARGE WASTE-EATERS
(Large Detritivores)

SMALL WASTE-EATERS
(Small Detritivores)

SOIL, MINERALS, & WATER

This chart outlines the complex cycle of food, food-eaters, and waste matter that makes up the "food web." Plants supply food for plant-eating animals, or herbivores—a large group of animals ranging from elephants to bees. In turn, most herbivores are food for the meat-eating animals, or carnivores. Some animals, such as racoons, bears, and humans, eat both plants and animals and are called omnivores. All animals produce waste in the form of feces (FEE-sees) and, eventually, as dead bodies. Plants also leave waste matter when they lose leaves, drop fruit, or die. Dogs, vultures, crows, fish, and other large carrion-eaters consume some of the waste—the dead animals, but most of the wastes provide food for smaller detritivores (de-TRY-te-vores). Billions of insects, worms, and centipedes, as well as fungi, tiny microbes, and other soil and water dwellers process billions of tons of organic matter every year. Minerals, chemical elements, and water all re-emerge as recycled nutrients to feed the plants and begin the cycle once again.

inside the sealed bottle, all the materials recycling naturally. Some of his early bottled ecosystems are still alive today!

Meanwhile, scientists in the Soviet Union were trying to build life-support systems for travel into space. They conducted several experiments by sealing people up in small buildings with artificial light for weeks at a time. But in these experiments, known as Bios-3, only half of the food was grown inside the temporary greenhouses. The rest was brought in from outside, and waste material was disposed of outside.

The results were encouraging, but not good enough. Before humans would be able to live on Mars (and beyond) they would have to build complete life-support systems on a much larger scale, a living system more like Earth.

The Soviet scientists inside Bios-3 harvested a hearty crop of grains under heat lamps and without soil, growing them in solutions of liquid nutrients. They also raised many vegetables. *Below:* A Soviet scientist takes his own blood pressure. Closed for up to six months at a time, Bios-3 was at one time the most advanced self-contained system in the world.

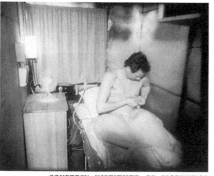

COURTESY INSTITUTE OF BIOPHYSICS

ANATOLI BELONOGOV/*SOVIET LIFE*

A few years ago, a group of people began to think about this challenge—how to make a replica of Earth so completely sealed off from our planet's air, water, and soil systems that it would have its own separate life cycles. They knew that creating a biosphere separate from Earth was a big undertaking. Could it be done?

They talked to Dr. Folsome in Hawaii and to the Soviet scientists. They met with the scientists at the National Aeronautics and Space Administration (NASA) who had sent men to the Moon. They consulted many other scientists, too, as well as engineers, architects, and builders.

Some people said no—they didn't think anyone on Earth was ready to try such a difficult project. But others said yes—it would take a lot of hard work, but it could be done. And that's how it began, with a small group of people willing to work very hard to make their dream come true, a dream that these pioneers named Biosphere 2.

Russian scientist Vladimir Vernadsky (1863-1945) was the first to present the idea that Earth is one huge, interrelated system of life which adjusts and sustains itself. For this valuable contribution to science he is considered the "Father of Biospherics."

19

GONZALO ARCILA

C. ALLAN MORGAN

C. ALLAN MORGAN

D. P. SNYDER

JOHN CANCALOSI

Surrounding a bird's eye view of Biosphere 2 are several of the Glass Ark's inhabitants. *Clockwise from top:* A four-eye butterfly fish, with the large black spot, and a French angelfish look for a meal on the coral reef; a karoo rose blooms in the desert; the pygmy goat, which lives in the Human Habitat next to the farm; the lush tropical orchard, which provides plentiful fruit, such as these papayas; the Jamaican anole lizard, which lives in the rainforest. Scientists had to anticipate the effects of each of these species on its fellow "biospherians."

CHAPTER 2

Building the Glass Ark

The first step in starting this historic experiment was to put together a design team. Experts in science, agriculture, architecture, and engineering were selected from all over the world. Men and women from many different countries and cultures joined together to make Biosphere 2 an international project.

The designers wanted Biosphere 2 to include a diversity of species: as many different plants and animals as possible. This is because some plants and animals will probably become extinct, just as they do in Biosphere 1. A greater diversity would mean a larger group of survivors and a better chance to keep the life cycles in Biosphere 2 going. They also had to be sure that each species had a large enough number of members to help it survive.

Biosphere 2 was also designed to have a variety of environments, just like Earth. There are five wild zones: a rainforest, a savannah, a marsh, an ocean, and a desert. There are also a "microcity" and a farm for the humans.

At first, Biosphere 2 was just a bunch of circles on a sheet of paper. "Here's the rainforest," said the planners. "That should probably be the tallest part. Here's where the desert should be. That should probably be the lowest part."

The planners soon realized that many things had to be included in the design to make Biosphere 2 as much like Earth as possible. They

MARIE ALLEN

Earth sustains a variety of environments, distinguished by temperature and rainfall. Ecosystems from Earth's temperate zones, such as this alpine meadow, cannot be included in Biosphere 2's tropical environment because their plant and animal life are adapted to cold winters and short summers.

21

The planning stages of Biosphere 2 included sketches, floor plans, and models. On the left, architect Philip Hawes works on a floor plan while rainforest scientist Dr. Ghillean Prance, Director of Kew Gardens in England, advises on the three-dimensional model. Hawes designed the biosphere's structure along with Margret Augustine, the project's chief executive.

needed a roof with varying heights to let the air rise and fall as it does in nature. They also needed to separate the rainforest and the desert as much as possible, so the jungle plants wouldn't invade the desert or the desert plants invade the jungle.

Since life inside Biosphere 2 would be dependent on the sun, the structure needed thousands of glass windows to let in the sunlight. A steel framework was required to hold up all the heavy glass. And a steel bottom was needed to separate Biosphere 2 from the ground below.

The basic building block is a five-foot-long tube of steel called a spaceframe. Spaceframes are very strong but don't weigh a lot. They can be bolted together to make a building of any size or shape.

Glass window panes are mounted on the frames. Each pane weighs about 250 pounds! It took a crew of three men and a huge crane to lift each pane into place. This process is called glazing. A special sealing material of liquid silicone was put between the glass pane and the frame to keep air from leaking out. That was one of the hardest problems to solve, to make a sealant that would be airtight.

Another big problem was air pressure. As the sun heats up the air inside Biosphere 2 on hot days, that air expands. In the summer, the sun could expand the air so much that the roof would blow off! The solution to this problem was a pair of giant-sized lungs connected to the main structure by underground tunnels. The lungs expand and contract as the air heats up and cools off. This prevents the glass from exploding or collapsing.

JOHN CANCALOSI

C. ALLAN MORGAN

MARIE ALLEN

C. ALLAN MORGAN   C. ALLAN MORGAN

Clockwise from top left: A section of roof, assembled on the ground, is lifted into place by a crane; secured by safety belts, workers then bolt the spaceframes in place; one of 6,400 glass panes is lowered by crane while the construction crew directs its positioning; the dome sheltering one of the two lungs nears completion; with the help of climbing ropes, clamps, and a sharp eye for safety at all times, workers scale the glass mountain to seal the windows into their airtight frames.

This aerial photograph reveals the complexity and size of the project. Surrounding the Glass Ark are dozens of "support" buildings. Near the top of the picture are the greenhouses, the Test Module, the laboratories, and the offices. At top middle is the Insectary which houses thousands of insects. On the left, Mission Control faces the already-completed rainforest and ocean areas. In this building scientists and support crews keep a constant watch over the biosphere and its inhabitants.

Gradually, the designers' rough sketches became detailed plans. At the same time, the search began for a very special place, the home of the Glass Ark.

A glass roof and glass walls didn't guarantee that Biosphere 2 would get enough sunlight. It also had to be in a sunny spot. But there are many such places on Earth. Which would be best?

The choice was narrowed down to somewhere in the United States because the United States is the easiest country in which to get the permits needed to build and to fill such an unusual greenhouse. It also has one of the best scientific communities in the world, with experts who could help work on the project.

The best place in the United States for sunlight and mild winters is the sun-drenched Southwest. The final choice was a beautiful ranch in the foothills of the Santa Catalina

Mountains about thirty miles north of Tucson, Arizona. Most people agree that it's a breathtaking spot. It also gets plenty of sun year around, but is high enough in elevation to miss the really scorching summer heat of the lower desert.

In 1987, a groundbreaking ceremony was held and the construction began in earnest. Biosphere 2 was a dream no more.

Troubleshooting in the Test Module

Even before the first closure of Biosphere 2, its famous Test Module caused quite a stir. This miniversion of the real thing was absolutely crucial to the project. It was here that the first experiments were made to see if everything would work in Biosphere 2 as expected.

First the planners had to be sure that the Test Module was leakproof. Then it was time to test the plants. They were sealed in the module for a month and given the same high humidity and light levels they would receive in Biosphere 2. How would they do? The answer would be in the first whiff of air when the module was opened. If the different gases got out of control, the air would be sure to smell funny.

The opening of the Test Module at the end of the month was a happy event. The sweet air and the healthy growth were very good

The Test Module glows with an eerie light at dusk as the lights of Tucson twinkle in the background. The module was built to be a testing device for the design, materials, and operation of the entire project. Biosphere 2 is over 400 times larger than the little Test Module, whose volume of 17,461 cubic feet is dwarfed by its giant neighbor's volume of 7,205,737 cubic feet.

25

Early tests in the Test Module convinced planners that Biosphere 2 could be a success. *Clockwise from upper left:* The module's tiny apartment has a kitchen, bathroom, bed, dining table, and desk; biospherian candidate Norberto Alvarez checks on "Vertebrate X," John Allen, originally trained in mining metallurgy and executive management and now director of research and development at Biosphere 2; during her stay in the module, biospherian Abigail Alling (Vertebrate Y) cares for the plants, chosen for their nutritional value and ease in growing.

grew at an astonishing rate, proving that they had achieved not only a working environment, but a successful, nourishing one.

Next came tests with animals. The first was to see if bees could navigate properly inside the glass building without the sun's ultraviolet light to guide them. They did. Then came the famous "taste test" for termites. Termites play an important role in soil recycling, but they eat almost anything. The sealant holding the glass windows to the spaceframes might be a problem. What if the termites ate that?

To find out the answer, several species of termites were fed "sandwiches" with their favorite food on the inside and the sealant on the outside. In order to get to their food, they had to eat through the sealant. Fortunately, none of the termites ate the sealant.

Perhaps the most exciting tests of all were those involving Vertebrates X, Y, and Z. These were the code names for the three people in the first human experiments.

The first test in 1988 was with one of the creators of Biosphere 2, John Allen. Several Soviet scientists had lived in partially closed systems for long periods of time. But John Allen would become the first person on Earth to live in a completely closed ecological system.

For three days, "Vertebrate X" lived inside the Test Module. His air was recycled, along with his water and waste material. His food came from a miniature tropical garden inside, which included everything from pineapples to potatoes and tomatoes to tea. Outside, a doctor and crew of scientists monitored his condition around the clock. At the end of the three days, John Allen emerged in excellent shape.

Just as important, the Test Module itself was doing fine. The air was good and the waste-recycling system was a success. The marsh plants were growing into a jungle.

Longer tests followed. Vertebrate Y, marine biologist Abigail Alling, spent five successful days in the Test Module. Alling noticed changes in her own body during her stay in the module. Most noticeable was the absorption of the miniclimate's humidity into her skin. But Abigail Alling felt that she had not been in the module long enough for the carbon dioxide/oxygen cycle to reach a steady balance in the enclosed environment.

Vertebrate Z, botanist Linda Leigh, underwent a much longer test. She lived and worked in the Test Module for three weeks. Like Allen and Alling before her, she was also keenly aware of the effects of her actions on the module's environment. In this experiment, however, Linda Leigh was inside long enough for the cycles to stabilize. She learned to do her daily chores in harmony with the rhythms of her closed world. For example, mammals release larger doses of carbon dioxide when they exert themselves. So she learned to dig up her sweet potatoes for dinner during daylight hours, allowing the plants time to absorb the buildup of carbon dioxide.

She also enjoyed exercising her imagination while she lived in this special world. She said later, "I was pretending that maybe it was a Martian or lunar terrain outside the module." She also said that what impressed her most was how much "everything I did had an influence." She came to realize that her actions affected all of her surroundings—from the plants and the insects, to the water and air. This cause-and-effect relationship is not usually so obvious in Biosphere 1.

Vertebrate Z, Linda Leigh, takes a sample of her own blood for testing during her trial run in the Test Module. Medical checks were an important part of all the human experiments. Today Leigh, like Alling, is part of the first crew inside Biosphere 2.

C. ALLAN MORGAN

Imitating Mother Nature

Creating a miniworld from scratch is no simple task, especially when nobody has done it before. First you must copy as closely as possible the balancing act that makes Earth work. Then you have to keep it going.

From the start, recycling was seen as the key to keeping the balance in the Glass Ark. Everything must be used and reused. Old elements must be constantly absorbed and changed into new elements with nothing going to waste. Life could not exist without this steady cycle of change that keeps everything in balance.

Originally, the planners hoped that solar power would provide the electricity inside Biosphere 2, just as the sun provides the power for Biosphere 1. But the equipment needed to do this was too expensive. Instead, generators make electricity for the lights, the computers, and all the mechanical devices inside.

There is also a separate energy system for heating and cooling the air. Many Biosphere 2 plants and animals are from the tropics, so the temperature must be kept above 55 degrees Fahrenheit in the rainforest and above 35 degrees (F) in the desert.

In order to maintain a healthy system, the air must be kept moving. It must follow a pattern as close to natural convection patterns as possible. In the Glass Ark, natural convection moves the hot air from the low desert to the higher rainforest. Pumps draw the air from the rainforest down through cooling coils in the basement where the air cools even more. Pumps and fans then help move the cooled air

C. ALLAN MORGAN

Biosphere 2 has its own electric power plant, shown here at dusk. Inside the cooling towers on the right are huge artificial waterfalls fed by ten million gallons of water per year. These towers help control the temperature in Biosphere 2 by sending chilled water through a vast system of underground pipes.

Top: The newly planted desert soaks up the bright sunlight of the Sonoran Desert. Temperatures over 100 degrees Fahrenheit are common here in the summer. *Bottom:* Less than 300 feet to the north of the desert is the rainforest, where mechanical devices help keep the temperature below 95 degrees (F) and regularly supply the lush tropical vegetation with mist and rain. In Biosphere 1 these two environments are usually many miles apart.

C. ALLAN MORGAN

GILL C. KENNY

30

back to the desert again. The small caves dotting the desert hide air shafts that lead to these pumps and fans.

The air must also be kept clean. In Biosphere 1, Earth's huge oceans and billions of tons of soil and plant matter absorb carbon dioxide and other harmful elements over long periods of time. Because Biosphere 2 is so much smaller than Earth, its atmosphere needed some help in cleaning its air.

So the planners developed "soil-bed reactors." A soil-bed reactor is a bed of soil through which air is pumped. The entire farm area has been designed as one huge soil-bed reactor; there are fans pushing air up through the planting fields all the time. It may sound strange, but the soil actually cleans the air!

Many people don't realize how important soil is. They think of it as something that holds plants down or dirties their clothes. But more importantly, soil is home to billions of busy little helpers: the mighty microbes.

Most of these tiny creatures can be seen only with a microscope. But no world can live without them. Microbes absorb both poisonous and non-poisonous elements from the air and change them into carbon dioxide, water, and

The soil is a living entity, filled with microbes like these, shown magnified hundreds of times their real size. *Left:* A red-encrusting algae. *Right:* A harpacticoid (har-PAC-te-koid), a microscopic insect that lives in water.

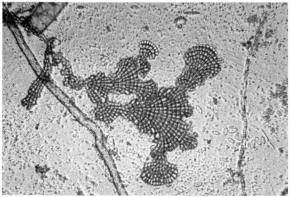

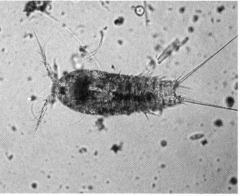

Top: The winding savannah stream, which has its own system of recycled water, had to be dug, lined with concrete and rocks, painted with a special acidic paint, and then planted with savannah plants. *Bottom:* By closure, this view of the savannah was full of grasses and trees.

other elements. They're the key to fresh air in Biosphere 2, just as they are in Biosphere 1. Right now the microbes underground, along with the insects, are busy cleaning the air all over the world. But no one breathing that clean air sees them do it!

Like the air supply, the water supply must keep moving. Animals and plants can't live without water. This life-giving cycle had to be carefully planned in the Glass Ark.

The first step in recreating the water cycle was to create weather. When rain is needed, the computers turn on a system of pumps and sprinklers. In addition to frequent misting in the rainforest and light rain in the other wild zones, there are many other sources of fresh water.

One very important source of this fresh water is the mountain in the rainforest. At the top of the mountain, a stream plunges down a waterfall into a pool and then flows out to the edge of the savannah. There some of the water goes underground to be pumped back up to the top of the mountain. The rest supplies a stream in the savannah that goes into a fresh-water pond at the edge of the marsh. The marsh water eventually ends up in the ocean, then goes through the desalination system down in the basement to remove salt, and is pumped back to the top of the mountain.

In addition to being important to the water system, the rainforest mountain is an excellent example of Biosphere 2's many "fakes." Made of concrete and plaster, it is a beautiful imitation of a real sandstone formation in Venezuela. Soil pockets were left in the sides to hold tropical plants, the same way they often grow in nature. And although the mountain now

Building an artificial mountain.
Top left: First, a skeleton of steel
rods was covered with a concrete
sealant several inches thick and
allowed to dry. Then another
layer of concrete and sand was
applied *(bottom left)* and carefully made into the proper shape
and texture *(bottom right).*
Finally, the whole thing was
painted charcoal grey *(middle).*

33

As in the other biomes, the desert was created on a foundation of carefully chosen soils and rocks. Here workers arrange the "real" boulders with the help of a bulldozer. The huge black rocks in the background are actually air shafts made to look like natural caves.

looks real, it's actually hollow inside with a spiral staircase leading to the basement.

There are other examples of imitations in Biosphere 2, created when the cost of hauling in the real thing was too great. The cliffs above the ocean and the boulders and caves in the desert are other amazing fakes made from special concrete on which real lichens can grow.

The inside weather forecast not only calls for showers, but also for changing seasons. Here was another challenge: how to be sure that it wouldn't be summer all over Biosphere 2 at once. That would cause a disaster. If all the plants grew at the same time, they might consume all the carbon dioxide and wind up dying of starvation. Then the animals would die because there would be no food or fresh oxygen.

The solution was to have some areas that bloom and grow in the winter and some areas that bloom and grow in the summer. The winter blooms are mostly in the desert.

During the summer, while the desert is "asleep," the savannah takes in large doses of carbon dioxide. Then the desert takes in its share of carbon dioxide during the winter while the savannah snoozes. The rainforest, marsh, ocean, and farm are always active.

A cooling system is also crucial in controlling the weather. On a hot summer day without a cooling system, the temperature inside could jump to 150 degrees (F). The inhabitants would be done for! As a safeguard, the cool air comes from two sources of chilled water. If one breaks down, the other can take its place. The heating system also has a backup, with two boilers instead of one.

These are the basic parts needed in Earth's first self-contained miniworld. Each part must work to keep it in balance. Each plant and animal inside must do the same, and every one of them is real!

This canna flower brings vibrant color to the rainforest.

LINNEA GENTRY

Earth supports an astonishing array of life. Over thirty million species of plants and animals have been identified so far. Scientists think there are many more not yet discovered. And although we don't yet fully understand the system, ecologists think that all these creatures are interrelated in our biosphere's complex web of life.

Top: A lion-faced macaque (ma-KAK) from southern Asia pauses while foraging for food. *Middle:* A small herd of white-tailed deer browse at the edge of the Biosphere 2 site. *Bottom:* These beautiful macaws are native to the tropical climates of Central and South America.

CHAPTER 3

All Aboard

As you can imagine, the planners faced a hard task in choosing the right mix of animals and plants to keep the system running smoothly. Noah had it easy compared to this!

Each of the zones, which are also called biomes (BY-omes), has its own captain. The captains were in charge of selecting and helping to install the plant and animal residents of their zones. Every captain started out with a "wish list" of favorite species. Then the lists were pruned for the good of the whole system.

The selection wasn't easy. From the start, only plants and animals with special qualifications were considered. Dangerous animals and plants were immediately eliminated. No poisonous snakes and no poison ivy. There were enough problems without creating new ones.

LINNEA GENTRY

Each animal had to contribute something to the food-web system, be easy to care for, and not weigh or eat too much. That left out elephants. It also nixed horses, whales, bears, and all other large animals. So most of the Biosphere 2 inhabitants are small.

Size considerations also affected the selection of plants. The brazil nut is a tasty treat, but its tree grows about a hundred and fifty feet high. It would have burst through the roof!

The tiny mouse deer was also on an early list. This pretty grazing antelope helps recycle

C. WALKER.
COURTESY EVERGLADES NATIONAL PARK

This Siberian tiger *(top)* and American alligator *(bottom)* are too large and require too much food to be inhabitants of Biosphere 2. The only large predators allowed inside the Glass Ark are humans.

When a hummingbird sticks its long beak in a flower, its head gets dusted with pollen from the top of the blossom. As it goes from flower to flower drinking nectar, it transfers the pollen to other flowers and helps fertilize the plants.

grass. However, it turns out to be too nervous for the job. If startled, it might throw itself off the cliff. Dr. Ghillean Prance, the captain of the rainforest, had hoped to have a monkey. But they eat too much.

So—no brazil nut, no antelope, and no monkey. Those species which were chosen are very important, each in its own way. The hummingbird is a good example.

Most hummingbirds are tough and fearless, so they have a good chance of surviving the experiment. Even more important to the Glass Ark, the hummingbird is a pollinator. Its special job is carrying pollen from flower to flower so plants can reproduce.

There are plenty of hummingbirds to choose from: 340 species to be exact. The final candidate had to fit a very demanding list of qualifications.

First of all, Biosphere 2 needed a hummingbird that didn't migrate too far. Otherwise, it could wind up crashing into the glass. It also needed to be a hummer that wouldn't fly too high. Some hummingbirds have a fancy mating dance in which they zoom up 100 feet into the sky. Too dangerous with a ceiling overhead!

Next, the scientists had to consider the birds' beaks. All hummingbirds have long, slender bills for sucking nectar from flowers. But they come in many different shapes. For Biosphere 2's wide range of flowers, a hummer with a straight beak about 20 millimeters (8/10 of an inch) long was required. The pretty sapphire-spangled hummingbird fit the bill.

Now the scientists faced another big question. Could they keep the hummingbirds alive?

THE LEGEND OF THE HUMMINGBIRD

The hummingbird plays an important role in another version of the story about the Ark and the Great Flood of long ago.

It is a legend handed down to us by the Pima Indians of Arizona. In their story, too, many people and animals took shelter on a big boat. After the rains stopped, Spider Woman sent out the fearless hummingbird to see if any dry land had appeared. When the hummingbird came back with a flower in its beak, the Pimas knew the waters had begun to dry up. After the boat reached dry land, the hummingbird gathered up some clay from the shore and took it to the Supreme Being. From this clay, the Supreme Being formed a new generation of human beings to repopulate the Earth. The Pimas called the new people "Children of the Hummingbird." So it seems only fitting that the hummingbird is as important an inhabitant of the Glass Ark today as it was on the ark of the Pimas long ago.

A pair needs about 3,500 flowers a day to survive. (If you've ever seen hummingbirds whizzing by, you know how much energy they burn.) After many calculations, they finally decided that Biosphere 2 could support one pair of hummingbirds. To play it safe, there is a backup supply of food for them.

This long-nosed bat is drinking nectar from a cluster of agave blossoms. Like the hummingbird, it will carry the pollen to other agave flowers. Many desert plants depend on the presence of bats to help their reproduction process. Farmers also count on bats to eat insects that might destroy their crops.

The hummingbird is not the only pollinator on the Glass Ark's maiden voyage. There are many kinds of pollinating bees, including stingless bees, carpenter and leaf-cutter bees, and Italian honeybees. Though the honeybees may produce enough honey to give the crew a treat once in a while, the bees will probably need most of the honey just to feed themselves.

Bats, moths, and other insects also help pollinate all the plants. Butterflies do more than pollinate. They bring beauty to Biosphere 2, something that the planners felt was very important. Also among the free-roamers are two small finches that eat mostly seeds.

While the hummingbirds and other daytime pollinators sleep, the bats are on the night shift, drinking nectar and hunting for moths. They eat about ten to fifteen moths a night apiece. But the odd thing is, each bat must try to snatch about 100 moths a night just to get those ten. Multiply that by the six bats in Biosphere 2, and you'll see that at least 600 "moth encounters" are required per night to support a small bat colony. Before the bats were approved to come aboard, the scientists had to figure out if there would be enough moths to go around. Fortunately, there were.

GEORGE LAWTON

During their walks through the wilderness areas, crew members may meet one of these black-necked garter snakes. These harmless reptiles, at home in several habitats, enjoy basking in the sun along rocky streams. They eat frogs, toads, and tadpoles. The name "garter snake" comes from their resemblance to old-fashioned garters!

Then there are the land-loving vertebrates (animals with backbones): some 40 kinds in all. Most of them are reptiles, including snakes, lizards, and turtles. The crew sometimes spots a blue-tongued skink, a yellow-footed tortoise, or the black-necked garter snake when they're out making their rounds of the wilderness areas. There are also plenty of amphibians inside, including the red-spotted toad, the southern spadefoot toad, and the canyon treefrog.

The mammals participating in this first closure are the leaf-nosed bats and the cute little galagos (GAL-eh-goes), also known as bushbabies. The galagos are very shy, but friendly when they get to know you. They are native to tropical rainforests and savannahs. Until arriving in Arizona, the three Biosphere 2 galagos had spent their whole lives in a university research laboratory! Now the rainforest is theirs.

MARGARET COLLINS

LINNEA GENTRY

Insects are a constant source of interest and of great value to us as well. Some of the insects inside Biosphere 2 include the carpenter bee *(top left)* and the ant *(bottom left),* shown here eating a termite. *Right:* A damselfly pauses above the pool in the rainforest.

The crew is particularly delighted to have them in Biosphere 2 and gives the good-natured galagos plenty of attention and companionship.

From the more than 700,000 species of insects on Earth, about 50 species were chosen to live inside the Ark. That doesn't include stowaway insects that may have snuck in during construction. Besides the pollinators, there will be ground feeders that break down old fruit and dead plants and animals into forms that can be used by other organisms. Ants, termites, and cockroaches are very important residents here for this very reason. Yes, cockroaches! There are three species of cockroaches, fourteen species of ants, and seven species of termites inside Biosphere 2.

There are also water insects like whirlygig beetles that live on the surface of the pond, pest controllers like the ladybugs that live in the farm area, and many insects that are food for other animals inside. Reptiles and amphibians eat a lot of insects!

The animals are only part of this story. Another leading role goes to the plants, of course. Not only are they important in keeping

The wide-eyed, nocturnal galagos have ears that can rotate in search of sounds. Although born in captivity, the Glass Ark's three galagos are free to travel through the rainforest on the network of tree branches and vines.

the balance of atmosphere and soils healthy, they're also food for the animals and humans.

Some Biosphere 2 plants can be used for medicines, too. Almost a quarter of all medicines in the United States use ingredients made from tropical plants found in rainforests.

Medicinal plants also grow in savannahs and deserts. But the ones in Biosphere 2 probably won't be used to produce medicine during this first closure.

43

Top: The rainforest environment along the Orinoco River in Venezuela, which the Biosphere 2 rainforest was designed to resemble. The Amazon and Orinoco rainforests of South America are crucial in recycling the world's excess carbon dioxide. *Right:* Biosphere 2's rainforest also provides the same service to its miniature world. Here it appears lush with a year's growth.

Other plants like the agave (ah-GA-vee) in the desert or the panama hat plant in the rainforest produce fibers for ropes and baskets. The desert jojoba (ho-HO-ba) produces oil and the rubber tree produces a milky-white latex. Some plants even produce soaps and gum.

Every inhabitant of Biosphere 2 had to be tracked down and brought back alive. Many plants and animals came from research centers, botanic gardens, plant nurseries, and other "tame" sources. In other cases, scientists had to hunt for them out in the wild.

While the species selection continued, other events were also going on. Little by little, the new homes for the future inhabitants were starting to take shape.

Welcome to the Wild Zones

Every part of the Glass Ark was a real challenge to make. This was especially true of the wilderness areas.

At the north end of the huge structure is the rainforest, most of it an imitation of the Amazon jungle. Rainforests are Biosphere 1's most richly packed areas. They swarm with life. But many rainforests are being destroyed by people who don't respect the land and want it for their own purposes.

At the same time, other people are fighting to save the rainforests. The creators of Biosphere 2 hope that their miniature rainforest will help ecologists learn new ways to restore areas of rainforest that have been destroyed.

HARRY SCOTT

Biosphere 2's planners sent an expedition to the Venezuelan jungle to select possible candidates for their half-acre of rainforest. Here biospherian Linda Leigh examines some tropical moss. Everything brought back to the Biosphere 2 site was isolated for months in special quarantine areas to make sure no pests or diseases would contaminate the other candidates.

Right: This is the "cloud forest" at the top of the mountain. The small plants and shrubs are adapted to fog, rain, and soggy soil. *Left:* The White's treefrog is a common resident on and around the mountain. Like most other tree frogs, this amphibian's toes expand into sticky pads that are useful for climbing on wet surfaces.

Here inside Biosphere 2, the 50-foot-high mountain is topped with a cloud forest of mosses, orchids, and bromeliads (bro-MIL-ee-ads). The leaves of the bromeliads form bowls that capture rain and make little ponds for insects, frogs, and geckoes to live in. Orchids, tree ferns, trees, and vines grow down the side of the mountain. Peach palms, papaya trees, and the panama hat plant grow down below, along with the tallest plants in Biosphere 2, the ceiba (SAY-bah) and rubber trees.

At the edge of the rainforest, sun-loving plants help protect the interior forest from the scorching desert sun. These include ginger shrubs and banana trees, a favorite of the bushbabies and the bats. There are many other kinds of fruits, fragrant spices, and flowers here, too.

There are lots of frogs and snakes and insects here, including mosquitoes, an important source of food for many wild animals. There are also plenty of snails and earthworms.

Savannahs are the grasslands of the tropical world. In Biosphere 1, they are usually found between deserts and rainforests and have trees and shrubs spaced far apart with wide expanses of grass between them. Most people know savannahs as the homes of some of the most spectacular mammals left on Earth. Elephants, giraffes, zebras, cheetahs, and lions all live on African savannahs. Biosphere 2's savannah is a bit more tame.

It is located on a rock cliff overlooking the ocean. In this narrow space, there are about seventy-five varieties of grasses that came from

Dr. Peter Warshall (in the red shirt) was the captain of the savannah area, as well as advisor on the vertebrate animals for Biosphere 2. While on a collecting expedition in the Rupunini savannah in South America, he checked out the elegant mound of a termite colony. These termites were only one of several species considered for the Glass Ark from among the 2,000 known species of termites around the world.

47

Africa, South America, and Australia. Most were chosen for their seed size, because the seeds had to be big enough for the finches to eat.

This is the home of Biosphere 2's main grazing animals, the termites. You may be surprised to hear termites called "grazing animals." Each year, old grasses on the savannah must be broken down to make their nutrients, or valuable food parts, available for next year's growth. In Biosphere 1, fires and large grazing animals (like buffalo and antelope) help this process along. They break down dead grass so that the microbes can eat them and recycle the nutrients back into the soil.

But fires aren't allowed in Biosphere 2 and there's not enough room for large grazing animals. So termites, Earth's first-class recyclers, are more vital than ever here. They're the key animals of the savannah.

The spookiest part of the wilderness might be the marsh. You never know what may jump out at you in a place like this! The plants crowd closely together, so it's difficult to see more than a few feet ahead of you. And there is very little solid ground. Dark, slow-moving water covers almost everything and the air is warm, moist, and heavy.

This marsh is known as an estuary, because it's fed by both salt water and fresh water. In Biosphere 1, estuaries are formed where the fresh water from rivers meets salt water from the sea. The mixture of salt and fresh water provides a rich supply of food for all kinds of plants and animals.

This estuary is a replica of the Florida Everglades. Of course, there are some big differences. For one thing, there are no deadly water moccasins or awesome alligators here!

Oysters and crabs, most of them native to the Caribbean, are being carefully introduced into Biosphere 2's marsh. The dense tangle of mangrove roots give shelter to both land and water creatures. As in Biosphere 1, productive wetlands such as this marsh are vital to the whole environment.

D. P. SNYDER

48

But there are many similarities in the drone of the insects, the croaks of the frogs, and the sheltering canopy of the mangrove trees.

The mangrove trees were transported by truck from Florida. The black and white mangroves are separated from the red mangroves by the oyster bay, where oysters, mussels, and clams live. Crabs are especially plentiful in the red mangrove thicket, which is the saltiest part of the marsh. There are also salt-tolerant plants here, known as halophytes (HAL-oh-fites). They're fed to the animals on the farm and the crew sometimes eats them in salads.

From the marsh, the water flows into the miniature ocean on the east side of the biosphere. Dr. Walter Adey, a marine biologist from the Smithsonian Marine Systems Laboratory, and the Biosphere 2 staff headed by Abigail Alling succeeded in creating the largest artificial marsh and ocean system in the world. The water reaches a depth of 25 feet and is

The marsh gradually changes from fresh to salty water until it flows into the ocean, which takes up approximately 15 percent of the total surface area of Biosphere 2's wilderness. Just behind the low wall at the edge of the marsh is the long trough that releases the waves that keep the ocean water moving. Down at the opposite end of the ocean, the rainforest looms over the cliff next to the sandy strip of beach.

home to a living coral reef and hundreds of other marine inhabitants.

A coral reef is a world in itself. These magical communities are the most densely populated places in Biosphere 1's oceans. Built by tiny marine animals called polyps, they bustle with constant activity. You can find such creatures as the clownfish hiding in the anemones (ah-NEM-oh-nees) and the cleaner wrasse (RASS, rhymes with brass) cleaning other fish of parasites.

The coral reef came from the Caribbean Sea and the Gulf of Mexico. It has about thirty different kinds of coral in it and almost 1,000 different species of animals living on or near it.

If you ever visit Biosphere 2, you'll be able to see this beautiful world through underwater windows set into the wall of the viewing gallery.

Can you imagine bringing the ocean to Arizona? Some of the ocean water came from the

Left: A diver in the Caribbean Sea, off the coast of Mexico, hunts for candidates to bring back to Biosphere 2. The handsome sea turtle is too large a predator for Biosphere 2's miniature ocean. *Right:* The coral reef in Biosphere 2 is home to many colorful and exotic fish, such as these four-eye butterfly fish, little blue chromos, and the porkfish with the dramatic black bands across its head.

GONZALO ARCILA

GONZALO ARCILA

GONZALO ARCILA

Biosphere 2 divers collect several species of coral at a government-approved site off the coast of Mexico. Among the many species brought back to the Glass Ark are sea fans, staghorn coral, cactus coral, brain coral, and fire coral. Corals are actually tiny animals, or polyps, that live in huge colonies and are most active at night. Biosphere 2's ocean water must be kept moving at all times so that nutrients and floating debris don't smother or poison the sensitive coral polyps.

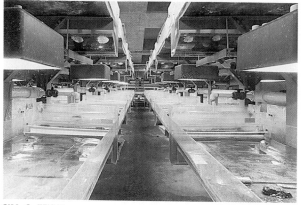

In the mechanical algae-scrubber system, these troughs of algae feed on waste and other tiny particles in the seawater. In this way they "scrub away" excess nutrients. They also add oxygen to Biosphere 2's ocean in the same way waves and tides do in Biosphere 1's ocean.

Pacific Ocean near San Diego, California. Over 100,000 gallons were hauled in by thirty-eight trucks. Believe it or not, the rest of the ocean comes from mixing up a salt recipe called "Instant Ocean."

It's crucial that the salt water around the reef be kept clean and undiluted. Otherwise the polyps could die. Here, the ocean is assisted by the "algae scrubber" system. Tiny green algae growing on screens eat the excess nutrients and bacteria from the water. Sixty of these "algae scrubbers" are located in a room under the savannah. The ocean water and the marsh water pass through them constantly.

Unlike a real ocean, there are no dolphins or sharks here. Dolphins eat too much and most sharks are too dangerous. But there are starfish, angelfish, sea urchins, and seaweed.

Biosphere 2's ocean also has waves and tides. Waves help circulate valuable nutrients that would otherwise pile up at the bottom of the sea. A special wave generator makes small waves that wash up on the beach about every twenty seconds.

Tides are also very important to the estuary. They bring the salt water to the marshes. Pumps help raise the water level when it's time for a high tide. Gates then control the flow of water into the marsh.

Although it's small, the beach has wind, waves, tides, and seashells, everything to make the crew feel right at home.

Some people think that the desert is the most enchanted place of all. It's a quiet corner in a world otherwise crammed with life. There is more space here, more breathing room. Since there is less water and fewer nutrients in the soil, the plants don't crowd together so closely. Each one stands somewhat apart, emphasizing the beauty of its flower—as in the brilliant blossom of the barrel cactus—or the strangeness of its shape. One of the strangest desert plants here is the boojum (BOO-jumb) tree, which looks like a huge, upside down carrot with toothpicks stuck in it.

You may be surprised to learn that there are many different kinds of deserts. Most of them are homes to lots of wonderful, hardy plants and animals. The desert in Biosphere 2 was designed by botanist Dr. Tony Burgess and is modeled after a cool, coastal desert such as those along the coast of Chile in South Amer-

Left: Several boojum trees, like these standing tall in Baja California, Mexico, were planted on the slopes of Biosphere 2's desert. Described by one naturalist as the "gateway into a wizard's garden," the boojum sprouts small round leaves after rainstorms. *Below:* A spiney lizard awaits introduction into its new home.

LINDA LEIGH   C. ALLAN MORGAN

Although natural deserts are an important feature of Biosphere 1, they are expanding at what some scientists consider an alarming rate. Nevertheless, they are often places of beauty. Here a barrel cactus seems to glow with its own soft light against a "rock" wall in Biosphere 2. Its neighbor is an agave. Many plants here have the ability, as Dr. Tony Burgess says, "to switch from frantically exploiting rainfall to stubbornly surviving drought." This is a crucial talent of any desert survivor, plant or animal.

ica, along the southwestern coast of Africa, and in Baja California, Mexico. The wide array of plants here can all tolerate the high humidity and fog drifting down from Biosphere 2's rain-forest and ocean.

Temperature extremes here are not as dramatic as they are in Biosphere 1's deserts, however. That would have been too costly to reproduce. But the desert is both the coldest and the hottest place in Biosphere 2. The temperature sometimes drops to 35 degrees (F) on winter nights and climbs to as high as 110 degrees (F) on summer days.

D.P. SNYDER   C. ALLAN MORGAN

Left: The future members of the "Fearsome Thicket," shown here in pots, await planting in the thornscrub transition zone between the savannah and the desert. *Right:* The yellow-footed tortoise is one of several animals expected to roam this area frequently.

Just above the desert is the thornscrub forest, a transition zone between the savannah and the desert. Ecologists call such places ecotones. Dr. Burgess calls it the "Fearsome Thicket." With so many dense, thorny plants, it's easy to see why. Most of them came from Mexico and Madagascar. Animals like the tortoises spend a lot of time here. Their hard shells protect them from the thorns and cacti.

There's another transition zone between the rainforest and the savannah. It's called the gallery forest. The bushbabies spend a lot of time there. Living things from different sections interact in these "buffer zones," with the hardiest and most adaptable ones surviving, just like they do in Biosphere 1.

This cut-away drawing reveals the inside of Biosphere 2's Human Habitat. The dining room, kitchen, and food storage rooms form a separate wing between the Human Habitat and the wilderness.

CHAPTER 4

The Place for People

The eight crew members live in the building called the Human Habitat. It's actually an indoor village. The designers of Biosphere 2 had to create this "village" in a very small area. Naturally they wanted to keep as spacious a feeling as possible. It was part of their job to avoid any sensation of "cabin fever" in the crew's quarters inside their Glass Ark.

So the architects planned a two-story structure with several large rooms in contrast to the many small ones. In addition to the large central staircase winding up the tall central tower other side staircases offer alternate routes through the habitat. Long corridors hung with brightly colored wallhangings also add to the feeling of space.

An even more important feature of the habitat is its windows. There are windows everywhere—offering views out over the Arizona desert surrounding the biosphere, as well as views looking out over the farm and wilderness areas within the glass walls. The long views do a lot to keep the biospherians from feeling trapped.

The designers planned the crew's living quarters as carefully as the homes of the animals and plants. Each of the four men and four women has a separate apartment with a sitting room downstairs and a bedroom upstairs. The many windows allow the biospherians to enjoy

One of the Human Habitat apartments; the spiral staircase leads up to the bedroom.

potted plants throughout their "indoor" quarters. This abundant vegetation ensures plenty of fresh oxygen for both humans and farm animals at all times. No problems with polluted air in their neighborhood!

The apartments have bathrooms with showers and toilets that look like your bathroom at home. But that's where the similarity ends. Since every bit of water is precious in Biosphere 2, baths are not allowed—and showers must be kept short. Even more unusual is the fact that toilet paper is not allowed—there's no way to dispose of so much discarded paper within the delicately balanced ecosystem of Biosphere 2. Human waste can be recycled in a natural way relatively quickly—toilet paper cannot.

The human waste is recycled through a septic system, which has bacteria in it to break down the solids before the waste moves into

The planners of Biosphere 2 modelled their wastewater treatment after a system first developed by Dr. Bill Wolverton at NASA's Stennis SpaceCenter in Mississippi. Biosphere 2's minimarsh system is made up of a series of tanks which clean about 600 gallons of waste water per day. The system is being studied for use in normal homes as well as for use by city wastewater departments all over the world. Methods such as these offer safe and inexpensive treatment of organic and industrial wastes.

Co-captain Sally Silverstone runs some tests on the communication system soon after installation in the Command Center in the Human Habitat.

special mini-marshes in the basement. There the marsh plants soak up the rest of the waste. The water then goes on to the farm where it's used for irrigation. Run-off water goes back into the plumbing system.

Drinking water, however, comes from the moisture transpired (breathed out) by the plants. The moist air is sucked into special containers in the basement. There the water condenses and drips into stainless steel tanks, ready to be used again as fresh drinking water. Here, as in the other biomes, recycling is the key to success.

The habitat also contains an analytical laboratory for research, a medical clinic to monitor the crew's health, a recreation room, and an exercise room. A Command Center with communication equipment, a dining room, a kitchen, and a repair shop are also part of what its creators call a microcity. At the top of

Dr. Roy Walford inspects supplies and equipment in his Medical Laboratory. Dr. Walford will examine all members of the crew regularly to ensure they remain in good health and to watch for any unexpected changes in their physical conditions.

the white tower there's a library with lots of books and video tapes. It also has a telescope so the humans can study the stars at night.

It's cooler and less humid here than in the rest of the biosphere. The temperature is kept at a comfortable 75 degrees (F) and around 70 percent humidity.

Everyone shares the cooking and cleaning chores in the kitchen. All the crew members like to cook. In fact, cooking was a part of the biospherian training program for all the candidates. Even more importantly, all biospherians within the Glass Ark help to raise their food.

The Farm

The builders of Biosphere 2 call the farm the Intensive Agriculture Biome, or the IAB for short. It's located in the section farthest from the wilderness, right next to the Human Habitat. It's covered with nine curved arches that step down the slope on the southern side of the biosphere.

Jane Poynter was the biospherian in charge of developing the farm. Like many people in her native England, Jane loves to garden. But Biosphere 2's "garden" demanded more than an admirer of flowers. She studied intensive farming techniques for years in France, Puerto Rico, and Australia, as well as in the United States, before the first experiment for the Glass Ark even began. She had to learn all about pest control, soil preparation, natural fertilizers, plant selection, proper harvesting methods, and crop rotation.

Jane Poynter and her fellow biospherian candidates were assisted in this challenge to develop a successful farm in such a small space by Carl Hodges and his staff at the Environmental Research Laboratory at the University of Arizona. They had been studying new agricultural methods in unusual situations all over the world. Together these experts and the Biosphere 2 staff experimented with hundreds of potential crops in the research greenhouses.

Once construction of Biosphere 2 was well enough along, they moved their plantings under the giant canopy for more precise study during the two years before the first closure. They learned which plants grew best in the available sunlight of the Arizona climate, which take the least out of the soil during their growth, such as peanuts, and which plants can most successfully follow another in a particular plot. For example, you shouldn't plant beans in a plot that just grew oats; the oats release a chemical into the soil that beans can't tolerate.

Now the crew grows about 150 crops on the farm's half-acre, not all at the same time, of course. They rotate crops that yield the most

Left: Biospherians hope to raise three crops a year on each of the eighteen garden plots in the Intensive Agriculture Biome, or IAB, seen here as the Glass Ark is built around it. The rice paddies on the left are permanent; crops are systematically rotated in the other plots. This helps to prevent any one particular crop from depleting the nutrients in the soil. *Right:* No space is wasted in the farm's small tropical orchard. Harvests yield everything from familiar grapes and bananas to more exotic star fruit and kumquats.

C. ALLAN MORGAN

D. P. SNYDER

nutritious food in a small space. These include many kinds of beans, which are rich in protein and starch, and several different types of potatoes, which are easy to grow as well as nutritious. White potatoes can grow year around, but the tropical sweet potato can only grow in the biosphere during the intense daylight hours of summer.

Children on a visit before closure were intrigued by the many crops that already filled the farm. Here they look for fresh peas to sample. The harmless white spots were left on the leaves by salt in the irrigation water.

C. ALLAN MORGAN

Most of the fat in the crew's diet comes from peanuts, rather than from meat, such as most Americans are used to. Peanuts are pressed into a nourishing vegetable oil or made into peanut butter and spread on bread with jams made from their own fresh fruits. The leaves of the peanut plants also make good food for the pigs.

There is also less wheat than in a typical American diet. Oats, barley, and amaranth help make up for this lack. The crew grows lots of rice, as well. In general, the human diet in the Glass Ark is lower in animal protein and grains, but higher in vegetables and fruits than what most Americans eat at home.

The fruits include bananas, papayas, oranges, grapefruit, figs, strawberries, tropical apples, and guavas. Most desserts are made with fruits. And there are lots of different juices. Sugar cane supplies the crew's sweetener. There are also plenty of vegetables: everything from tomatoes to peppers to spinach.

As with any farm, the IAB has its share of crop pests. In Biosphere 1, these might be treated with chemical bug sprays. But that

C. ALLAN MORGAN

Biospherian Jane Poynter releases ladybugs into a plot rich with a grain called sorghum. These tiny beetles do a good job of warding off insect pests without damaging the plants.

could cause a disaster inside the Glass Ark. The Biosphere 2 world is so small that if someone sprayed a pesticide on a crop today, it could wind up in the drinking water tomorrow!

So the biospherians rely on natural helpers. Marigold plants keep some pests away with their smell. Ladybugs eat many others. Still others can just be squirted off the crops with water. The toxins that are released naturally by growing plants are just as naturally removed from the air by the soil-bed reactor system.

In many ways the IAB is a kind of pioneer space farm. Many of the crops are being studied for possible use in space stations and on other planets. One way the crops may get there is in tissue cultures, another important

Tissue cultures, in the form of single cells, take up much less space than seedlings and they last a much longer time. If a species suddenly dies out, biospherians can unfreeze a cell from the tissue culture library, place it in a test tube with the right nutrients, and plant it when ready. These test tubes hanging near a window are nurturing plants being tested for possible inclusion in Biosphere 2.

method by which plants in Biosphere 2 can be grown.

In this system, plants are produced by taking a cell from a parent plant and putting it in a test tube with some food. The cell multiplies and grows into a baby plant. When large enough, the young plant can be set in the ground. These single cells can be stored in much less space than plants grown in soil. They are also better than seeds, because they last in cold storage for a very long time.

When new plants are needed, humans go to the Tissue Culture Lab, pull the plant cells out of storage, and transform them into seedlings. This might be necessary if a species suddenly dies or develops an incurable disease.

Crew members are responsible for the feeding, breeding, and health care of the farm animals. These jungle fowl are descendants of the original chicken, native to India.

Animals in the House

Can you imagine a barnyard in your home? In Old World Europe, animals lived on the lower level of the farmer's house. They do the same in Biosphere 2.

Here, pigs, goats, and jungle chickens all keep each other company, looked after by biospherian Linda Leigh. The Asian jungle fowl is the original chicken from which all other chickens descended. They provide the humans with several dozen eggs a week, as well as meat. Although this breed lays fewer eggs than domestic commercial chickens, it is well adapted to the humid climate and takes good care of its chicks.

The African pygmy goats provide milk, cheese, yogurt, and butter. Once in a while they also provide meat. Full-sized American goats take up too much space and eat too

Biospherian candidate Silke Schneider from Germany trims a goat's hooves. Four does (females) and one billy (male) were taken into Biosphere 2. Milk from the goats will be the largest source of necessary fat in the crew's diet.

C. ALLAN MORGAN

much. As it turns out, pygmy goats are tougher and more independent. They also depend less on grain.

The pigs are also a pygmy species. They're a feral, or wild, pig that comes from the state of

In June of 1991 the farm animals were introduced to their new home inside the Glass Ark. The barnyard area is roomy and filled with sunlight. *Right:* Biospherian Jane Poynter tries to lure a reluctant sow (female pig) into her new home. A doughnut finally convinced her to emerge.
Bottom: A young visitor pats a goat newly released from its traveling cage; others are about to be lifted off a small cart.

D. P. SNYDER

D. P. SNYDER

Georgia. Biosphere 2 has one boar and two sows—and lots of piglets!

Just outside the Glass Ark, other candidates are being studied for future biosphere experiments. These include miniature sheep, which are only two feet tall!

Most of the domestic animals live in the village. But one of the most fascinating is being raised on the farm. That's the fish called tilapia (tee-LA-pee-ah). Tilapias come from the lakes of Africa and incubate their eggs in their mouths. They look like giant goldfish, breed rapidly, taste good, and help purify the water and fertilize the soil. In short, they're miracle fish!

In Biosphere 2, the crew is raising tilapia in the rice paddies. Here, the fish munch on algae and a water fern called azolla (ah-ZO-la). When the rice is harvested, the grain goes to the humans and the stems to the domestic animals. Meanwhile, the waste from the tilapias is converted into fertilizer for the rice crop.

Every few weeks, the crew has tilapia for dinner and the azolla is fed to the chickens. Does this unusual system work? It's been used in China for 4,000 years!

To increase the quantity and size of the fish, biospherian candidate Bernd Zabel experimented with and adjusted a fish-rice-azolla system that came from ancient China. This photograph of the tilapia fish was taken in the early stages of Zabel's project. The tilapia now grow in the rice paddies on the Biosphere 2 farm.

TYPICAL SUNDAY DINNER

Black Bean Soup with Freshly Baked Bread
Tossed Garden Salad
with Tomatoes, Cucumbers, Avocados,
and Nasturtium Flowers
Grilled Tilapia Fillets with Fresh Tomato
and Basil Sauce
Baked Sweet Potatoes
with Yogurt, Chives, and Parsley
Zucchini in Lemon and Oregano Sauce
Fruit Salad with Orange Sherbet

D. P. SNYDER

Here is the crew of the first Biosphere 2 two-year experiment. They are all people with highly developed skills and a wide array of accomplishments. *From left:* Mark Nelson, Jane Poynter *(top)*, Roy Walford *(bottom)*, Linda Leigh, Sally Silverstone, Taber MacCallum *(top)*, Abigail Alling *(bottom)*, Mark Van Thillo.

CHAPTER 5

The New Biospherians

Because this experiment is a trial version of future space colonies—which will be far away from the safety and comfort of Earth—the crew cannot go in and out of Biosphere 2. They will only be allowed out in case of an emergency, if someone gets very sick or if something spoils the atmosphere inside. You may well wonder what kind of people are living in such unusual circumstances and what kind of training they have had to handle the problems that such an unusual experiment may bring.

The human biospherians come from England (Jane Poynter and Sally Silverstone), Belgium (Mark Van Thillo, pronounced TEE-lo), and the United States. The Americans are Abigail Alling from Maine, Linda Leigh from Wisconsin, Taber MacCallum from New Mexico, Mark Nelson from New York, and Dr. Roy Walford from California.

These eight crew members were selected from a group of fourteen volunteers. They ranged in age from 27 to 67 years old and came from seven different countries. In addition to the three countries already mentioned, Germany, Mexico, Australia, and Yugoslavia were also represented. Most of the biospherian candidates had previous training in science or engineering and all of them have traveled over most of the world.

However, being a well-traveled scientist isn't enough. Crew members also have to be self-sufficient in everything from running tests in

ROBERT HAHN

Biospherian Taber MacCallum, Director of Biosphere 2's Analytical Systems, has studied chemistry, ecology, and computer science to prepare himself for his demanding job as the "chemistry watchdog" inside the Glass Ark. In Biosphere 2, state-of-the-art technology and ecology go hand in hand.

the labs to doing their own repairs. They are their own plumbers, electricians, and tailors, too.

Dr. Walford is a medical doctor, so he's able to make sure everyone stays healthy and to handle minor injuries and illnesses. All of the crew have taken courses in first aid treatment. Some crew members also took a course in dental care at the navy hospital in San Diego, California. But if someone becomes really sick, he or she will have to leave Biosphere 2 for medical attention in Tucson.

The biospherians are all physically fit and able to handle emergencies calmly and quickly. They've been trained to be good observers and researchers, as well as competent farmers and computer users.

During their extensive training, the human candidates took part in many scientific conferences to learn more about the biology, ecology, chemistry, atmospherics, and geology of our planet. They all helped search for soils, hunt for insects, and inspect farms all over the globe. They also know Biosphere 2 inside and out, since all of them helped build it.

In addition, most of the candidates went on numerous expeditions, many of them aboard a research ship called the *Heraclitus*. The *Heraclitus* has traveled the world's oceans to investigate the ecology of the seas, from the marine life along the Antarctic Coast to the tropical rainforests of the Amazon River. It was on board this floating research laboratory that biospherians Mark Van Thillo and Taber Mac-Callum not only learned about Earth's many ecosystems to prepare themselves for living inside Biosphere 2—they also became expert scuba divers. Mark, Taber, and marine biologist

JOHN HORNIBLOW

ISOLDE DROSCH

(Left): Biospherian candidates learn to work together as a crew while learning how to endure isolation and hard work aboard the *Heraclitus.* *(Above):* Shown here off the shore of Antarctica, the eighty-two-foot research vessel is named for an ancient Greek philosopher.

Abigail Alling all don their diving suits to care for the miniature ocean and its inhabitants inside Biosphere 2.

Biospherian Abigail Alling headed the marine research program aboard the *Heraclitus* for five years before she left the sea for the southwestern desert. Even before joining the *Heraclitus,* she had learned all she could about marine biology at Middlebury College, Yale University, and other institutions. She then spent many years studying whales in the Indian

JEFF TOPPING

C. ALLAN MORGAN   C. ALLAN MORGAN   D. P. SNYDER

From left: Abigail Alling (Maine, USA, 1959); Linda Leigh (Wisconsin, USA, 1951); Taber MacCallum (New Mexico, USA, 1964); Mark Nelson (New York, USA, 1947); Jane Poynter (Surrey, England, 1962); Sally Silverstone (London, England, 1955); Mark Van Thillo (Antwerp, Belgium, 1961); Roy Walford (California, USA, 1924).

Ocean for the World Wildlife Fund as well as other marine mammals around the world.

For Gaie, as she is called by her friends, all this experience undeniably helped prepare her to design, direct, and monitor the ocean and marsh biomes in Biosphere 2. Though she'll miss the "real" ocean, this new adventure is the most exciting phase yet of her exploration on the frontiers of science.

Linda Leigh grew up in the rich and diverse countryside of Wisconsin. Loving the outdoors, she naturally turned to the study of botany and ecology in college. She has worked on various projects all over North America, from replanting native prairie grass in the Midwest to studying wolves in the Northwest to range management in the Sonoran Desert.

She has had lots of adventures in her life as a field scientist. One involved an encounter with a nine-foot bear in Alaska! She said later, "I suddenly remembered what an old-timer had told me. When they charge, make yourself seem as big as you can." Linda rose up as tall as she could, spread out her parka, and successfully discouraged the huge bear from going through with his threat!

Taber MacCallum and Mark Van Thillo also have their stories about encounters with mysterious creatures down in the oceans and learning to overcome difficulties in wild places. Most importantly, all of the biospherians have learned that working as a team with a group of people you trust is the best preparation when you face the unknown. And right now, for them, Biosphere 2 is the ultimate unknown. It's Adventure with a capital A!

Daily life for the crew is much like it is on a ship. This first closure has two co-captains in charge: Sally Silverstone and Mark Van Thillo.

They check on everyone's schedule for the day at the morning meeting. Then the crew's daily routine begins. Everyone works on the farm planting and tending crops, making compost, and pulling weeds. There is also a lot of work preparing the food after harvest, like threshing the wheat and hulling the rice. There are pipes to fix, light bulbs to change, and other maintenance chores to do.

The person in charge of all the mechanical parts of Biosphere 2 is Mark Van Thillo. Mark is a very special "fix-it" man. He is a trained mechanic and tool-and-die maker with an

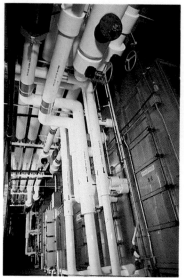

The basement in Biosphere 2 is a maze of water tanks, compost machines, drying racks, air chambers, electrical boxes, water pipes, control panels, condensation chambers, and other kinds of machines. These are just a few of the many pipes which require constant observation. Behind and under them are air handling chambers that help keep Biosphere 2's atmosphere moving.

expert knowledge of the life system that Biosphere 2's complex mechanical system must support. He knows what every fan, every pump, every pipe, every wire, every hose, and every motor is connected to inside—and outside—the Glass Ark. And he knows how to fix them.

Mark especially liked tinkering with electronic equipment as a youngster at home in his native city of Antwerp in Belgium. One time he unintentionally caused a blackout in his neighborhood! He learned more, however, at a technical college. He added to this knowledge by working at a busy oil-refining plant. Later he took charge of the transportation equipment for an ecology expedition in the jungles of Central America. His job as Chief Engineer aboard the *Heraclitus* was also important training for his responsibilities in Biosphere 2. And now, no matter what the problem is, no matter how puzzling a broken pump or a stripped gear may be, Mark usually manages to find the best way to fix it while keeping his sense of humor.

Aside from the responsibility of maintaining the huge Biosphere 2 mechanical system, the crew must also keep watch on the wild zones and step in to help when needed. Some scientists have described the humans as the "keystone predators." That's because they're the ones who control the balance of species in Biosphere 2 when necessary. For instance, if the population of a group of insects grows too large, the humans will have to eliminate some of them. Once in a while the humans fish in the ocean for lobsters and crabs, fulfilling another "predator" role.

The crew are stewards of the wild zones, as well as predators. Occasionally the humans put out extra food for the animals. And once in a while the crew weeds out some algae in the ocean. The algae scrubbers in the basement are cleaned regularly, too, a chore that all the biospherians help do.

Most of them spend their afternoons doing research, some in the labs and others in the wild zones. Everyone has their own specialty.

But life isn't all work. On the weekends everyone finds time to relax. While there are no baseball games in Biosphere 2, there are other activities to enjoy, like camping in the desert or swimming and snorkeling in the ocean. And the biospherians can keep in touch with their friends and families through telephone and video hookups. But no visitors are allowed inside!

Of course, there are televisions, video players, stereos, games, and a music room. Many crew members brought their musical instruments with them, so Biosphere 2 sometimes comes alive with the sounds of live music. Roy Walford brought his electronic keyboard. Taber

D. P. SNYDER

Crew member Mark Van Thillo enjoys some quiet time by himself, reading on Biosphere 2's beach. Other favorite pastimes in the biosphere are playing chess and listening to music.

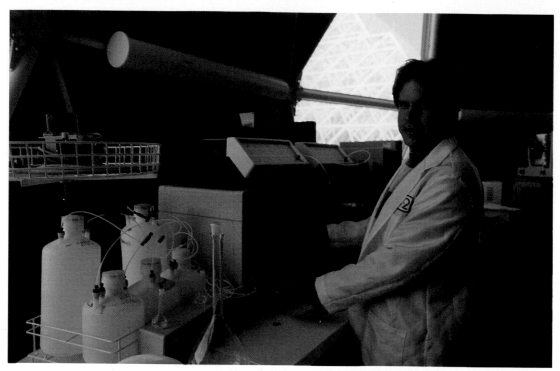

The sensing equipment monitoring air pressure, oxygen and carbon dioxide levels, temperature, and humidity must be checked several times each day. *Top:* Crew member Taber MacCallum begins adjusting new equipment in the Analytical Lab. *Bottom left:* Engineers inspect the recently-installed sensors in the farm area. *Bottom right:* Biospherian candidate Roy Malone Hodges measures salt content and temperature of the marsh before closure.

MacCallum brought along his cello and his drums. Linda Leigh brought her saxophone. And Jane Poynter has her balalaika. It's a wonderful way for a hard-working biospherian to relax after harvesting rice or oiling motors all day!

Checking the Pulse of Biosphere 2

Water, air, soil, plants, animals, and humans all must be watched and tested continually to make sure that the complex recycling system is working properly. The monitoring of Biosphere 2 was another challenge that the designers and engineers had to meet successfully.

This monitoring of the Glass Ark started long before it was closed and it continues constantly now. The crew patrols the various zones every day. They quickly spot any major changes, especially in natural sensors, like the giant clams. As long as the clams look healthy, the ocean ecosystem is probably fine. If they change color or shrink, that's a sign that the system has a problem.

Other monitoring is done by a nerve system of 2,000 electronic sensors that each take 360 readings every hour. The sensors are in key places, including inside the soil to keep track of its temperature. All the information is fed into computers in the Mission Control building, just outside Biosphere 2.

Since physicians and scientists sometimes need to watch people and other species on the inside to diagnose and solve problems, cameras are mounted in several places inside the Glass Ark. Visitors also like to see the inside more closely. They can watch the activities inside Biosphere 2 on television screens in both the visitors center and the viewing gallery beside the ocean.

The Analytical Lab is a major part of the monitoring system. The lives of everyone inside Biosphere 2 depend on how quickly and accurately the lab can check on the quality of their special world.

This means more than just checking to see if the air is still good to breathe and the water is fit to drink. For instance, if the carbon dioxide increases too much, the biosphere could get too hot. If the oxygen increases too much, spontaneous fires could break out. All the gases have to be in balance for its atmosphere to stay healthy.

If problems arise, the biospherians must rely on both their own judgement and on advice from experts outside. Computer linkups in the Command Room can relay the best advice possible to the biospherians in minutes.

Information is a two-way street. Biosphere 1 both learns from Biosphere 2 and offers advice to it. Someday, a nerve system like the one created for Biosphere 2 may be used to monitor parts of Biosphere 1 as well.

Norberto Alvarez, director of the Computer Center at Mission
Control, and his staff check information coming in from the
Biosphere 2 sensors, perhaps four times an hour. They will call up
ten-second readings, taken from continuous temperature sensors,
only if an unexpected change demands a closer look.

An artist's vision of what a future biosphere may be like at some cold and distant outpost. Here the glass and steel shelter a small farm, a playground and jogging path, a marsh, and a forest from the extremes of an arctic climate. Notice that the weak sunlight must be supplemented by artificial light. All future biospheres will have to fit the conditions of their location, whatever they might be.

CHAPTER 6

Visions of Tomorrow

At Biosphere 2, science fiction is rapidly turning into science fact. This exciting experiment has already started giving information back to its parent planet, Earth. More knowledge will emerge as scientists study this information.

New technology will be developed based on this new knowledge. This will probably include new ways to clean our air and water in Biosphere 1. The state-of-the-art monitoring system also promises improvements. All that we learn about managing our natural resources in Biosphere 2 is bound to be good for Biosphere 1.

Very likely there will be giant leaps forward in making tissue cultures. Think of it! Future space travelers may take the wealth of planet Earth with them in a cardboard box!

Some "pioneer" species may even evolve to adapt to the conditions inside Biosphere 2. These may even include the humans. If people are going to live elsewhere in space, they will have to adapt to the new conditions of isolation and small-group living awaiting them.

As for living on Earth, our thinking is sure to become more "planet conscious," as we discover how much every little part of life affects the whole. No insect, no mammal, no human stands alone.

But Biosphere 2 is just the beginning. New biospheres will surely be built. As new crews enter Biosphere 2 in the future, Biosphere 3 and

Launch of the Space Shuttle *Columbia*, one of the United States' fleet of reusable space vehicles. Many discoveries made in Biosphere 2 may be useful for future space flights.

NASA-COURTESY OF SPACE IMAGERY CENTER, UNIVERSITY OF ARIZONA

Biosphere 4 will probably be under construction somewhere. Inside these future "living laboratories" people will be able to study such mind-boggling problems as global warming, ozone depletion, and acid rain. They will do this by copying those conditions on a small, manageable scale. People will also continue to study individual species and how different plants and animals relate to each other.

The day draws nearer when humans will be able to pack their bags, microbes, tissue cultures, and animal companions, and set out for the space frontier. Or maybe future space colonists will send tissue cultures on ahead, along with robot builders. By the time people arrive, the biospheres will already be built and photosynthesis will be going strong! There are as many possibilities as you can imagine.

As you grow up, chances are that artificial biospheres will become familiar to you. You'll certainly hear more about them. Perhaps you will help build one someday—as a scientist, engineer, architect, or artist. Or perhaps you will be a biospherian yourself. Can you imagine living inside a glass ark like this for two years or more? Maybe you will someday live in a biosphere on Mars!

While nobody can say for sure what the future holds, we do know one thing: we decide our tomorrows by what we do today. Our dreams, our sketches, our research, and our models all help shape our future.

It's not so farfetched to think of yourself as a biospherian already. Actually, we are all biospherians. You, the plants, the animals—every living thing from the smallest microbe to the magnificent blue whales—we all belong to this special planet, Biosphere 1.

MARS SCIENCE STATION AND HOSPITALITY BASE

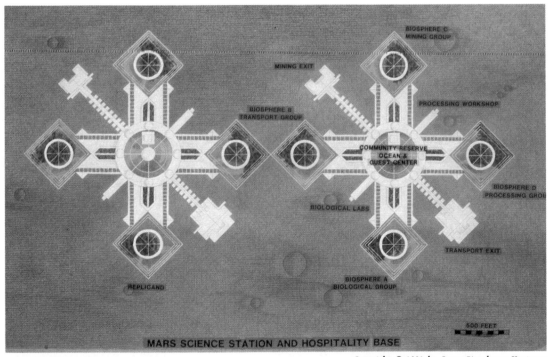

MARS SCIENCE STATION AND HOSPITALITY BASE

Scientists and engineers at Biosphere 2 are already thinking about how to build the first colonies on Mars. *Top:* A side view of a possible colony, with separate sections that look much like Biosphere 2. Note that workshops and an emergency shelter are deep underground. *Bottom:* The floor plan of such a colony, with a sister colony next to it. The main purposes of this imagined Martian base are mining and biological research. (Rendering by Sarbid, Ltd. Designers, Margret Augustine and Phil Hawes.)

For all its difficulties, Biosphere 2 is an experiment that cannot fail. Just to try it is a great leap forward. And just as Biosphere 1 isn't perfect, we can't expect Biosphere 2 to be perfect either. But this modern-day ark of glass and steel is sure to teach everyone a great deal about protecting our home planet and preparing for Mars and beyond.

And perhaps someday, far away from Earth, some other miniature biosphere may carry your children or your children's children off into space to bravely explore the hazards and wonders that await us there.

Who knows what marvelous discoveries the future will bring?

The outer door of the airlock chamber, before closure.

Someday other moons may be seen over other biospheres.

The Hard Facts About Biosphere 2

Location North Latitude: 32 degrees, 35 minutes
West Longitude: 110 degrees, 50 minutes
3900 feet (1190 meters) Elevation

	Square Feet	Square Meters	Acres
Total Airtight "Footprint"*	137,416	12,766	3.15
Footprints			
Intensive Agriculture	24,020	2,232	.55
Human Habitat	11,592	1,077	.27
Wilderness Area	62,590	5,815	1.44
Lungs	39,214	3,644	.36
Glass Surface	170,000	15,794	3.90

Dimensions	N/S x E/W**	N/S x E/W**	Height
Intensive Agriculture	136 x 177	41 x 54	80 ft.
Human Habitat	73 x 242	22 x 74	76 ft.
Rainforest	143 x 143	44 x 44	91 ft.
Savannah/Ocean	275 x 100	84 x 30	87 ft.
Desert	121 x 121	37 x 37	75 ft.
Marsh	91 x 63	28 x 19	8 ft. deep
Ocean	147 x 63	45 x 19	25 ft. deep
Lungs (each)	158 x 158	48 x 48	50 ft.

Volumes	Cubic Feet	Cubic Meters	
Intensive Agriculture	1,336,012	37,832	
Human Habitat	377,055	10,677	
Rainforest	1,225,053	34,690	
Savannah/Ocean	1,718,672	48,668	
Desert	778,399	22,042	
Lungs (at maximum expansion)	1,770,546	50,137	
Soil, Water, Structure,& Biomass	671,635	19,019	
Air	6,534,102	185,026	
Ocean water	133,690	3,786	1,000,000 gals.
Fresh water	26,738	757	200,000 gals. (approximately)

Living Biomass Approximately 70 tons (140,000 pounds)

Temperature Extremes Allowed	High	Low	(Degrees Fahrenheit)
Rainforest	95	55	
Savannah	100	55	
Desert	110	35	
Intensive Agriculture	85	55	

*Footprint = The area of ground covered by the sealed structure
**N/S x E/W = Measurement from the north side to the south side of that section x (by) the measurement from the east side to the west side of that section

For Further Reading

ECOLOGY AND ENVIRONMENT

Earth Works Group, The. *50 Simple Things Kids Can Do to Save the Earth*. Kansas City: Andrews and McMeel, 1990.

Lamber, Mark. *The Future for the Environment*. London, New York: Gloucester Press, Franklin Watts, 1987.

Lambert, David. (Isaac Asimov, editor). *Planet Earth 2000*. New York: Facts on File Publications, 1985.

Stevens, Lawrence. *Ecology Basics*. New Jersey: Prentice Hall, 1986.

SPACE SCIENCES

Bova, Ben. *Welcome to Moon Base*. New York: Ballantine Books, 1987.

Vogt, Gregory. *Space Stations*. New York: Franklin Watts, 1988.

ADVANCED

Allen, John. *Biosphere 2: The Human Experiment*. New York: Penguin Books, 1991.

Attenborough, David. *The Living Planet*. London: Colins, 1984.

Durrell, Lee. *State of the Ark*. Garden City, New York: Doubleday & Company, Inc., 1986.

Ehrlich, Paul R. *The Machinery of Nature*. New York: Touchstone Books, 1987.

Margulis, Lynn and Sagan, Dorian. *Biospheres from Earth to Space*. New Jersey: Enslow Publishers, 1989.

Snyder, Tango Parrish, editor. *The Biosphere Catalog*. London: Synergetic Press, 1985.

Glossary

Acid rain: rain containing poisonous amounts of sulphuric acid, which forms when the gases sulfur dioxide and nitrogen oxide are released into the atmosphere.

Algae: the simplest forms of green organisms, which usually grow in water. One example is the thin layer of green scum that forms on the sides of swimming pools.

Algae scrubber: a trough or tub that holds a removable pad on which algae live. The algae eat the tiny particles from the water flowing over the pads.

Anemones: a large group of sea creatures that attach themselves to rocks and have tentacles that look like flowers.

Atmosphere: the mass of air surrounding Earth; and any mass of gases surrounding a planet, star, or other heavenly body.

Bacteria: a large group of microscopic organisms with single-celled bodies, which live in soil, water, or in the bodies of plants and animals.

Biome: a habitat where plants and animals with similar characteristics live, such as a marsh or a prairie.

Biosphere: a complex group of living organisms that by living together help maintain their environment.

Botanist: a scientist who studies plants.

Bromeliads: a group of tropical plants, including pineapples, with stiff leaves that usually grow attached to trees and cliffs.

Carbon dioxide: one of Earth's most important gases, containing the carbon needed by all plants and animals.

Cell: the basic unit of life on Earth, made up of complex chemicals that can absorb food, grow, and reproduce.

Compost: a rich mixture made up mostly of decaying plant and animal material which is used for fertilizer once it is decomposed.

Convection: the movement of air whereby hot air rises and cold air sinks.

Coral reef: a rock-like growth consisting mostly of calcium made by coral polyps in warm seawater. It is used by many other creatures as a home.

Diagnose: to determine a disease or condition from its signs and symptoms.

Ecologist: a scientist who studies Earth's ecology.

Ecology: the delicate balance and interrelationship of air, water, soil, and life on Earth that makes our planet livable.

Ecosystem: a community of living organisms and their environment which form a functioning whole.

Ecotones: a transition zone between one ecological system and another.

Endangered species: a group of plants or animals whose population has become so low that it is in danger of becoming extinct.

Environment: the surrounding home of an organism or of a community.

Environmentalist: someone who studies and preserves Earth's environmental balance.

Estuary: the marshy area where a river or stream joins a large saltwater lake or the sea.

Evolution: the process of gradual change whereby living creatures become better adapted to the environment.

Extinct: completely died out; having no members of a species left.

Fungi: a large group of simple organisms that lacks chlorophyll, including molds (that you sometimes see on foods), mildews (that you sometimes see around bathtubs), and mushrooms.

Generator: a mechanical device that converts energy or fuel into electricity.

Glazing: the process of setting glass in a frame.

Global warming: the gradual rise in temperatures now going on throughout the world.

Halophytes: plants that can tolerate large amounts of salt in their soil and water.

Hulling: taking off the outer covering of a fruit, seed, or nut.

Incubation: the process of warming eggs to help them develop and hatch.

Keystone predators: the most important animals that prey on other animals and thus help to control their populations.

Latex: the milky sap of certain plants and trees which can be used in making such things as rubber and chewing gum.

Lichens: tiny lifeforms made of a combination of fungi and algae that grow in thin crusts on rocks, tree trunks, in soil, on walls, etc.

Life-support systems: mechanisms, whether natural or man-made, that enable living things to survive.

Marine biologist: a scientist who studies life in the sea.

Microbe: a very tiny form of life, usually a single-celled bacterium, that can be seen only with a microscope.

Natural resources: useful materials found in nature, such as wood, water, or minerals like copper and iron.

Nutrients: nourishing ingredients that are used as food by animals and plants.

Oxygen: a colorless, odorless gas that makes up 21 percent of Earth's atmosphere and is essential for plants and animals to live.

Ozone depletion: the loss of a layer of ozone gas in our atmosphere. Ozone blocks out dangerous ultraviolet rays from the sun. Scientists now believe this depletion of ozone has been caused by the chlorofluorocarbon chemicals used in air conditioners and aerosol spray cans.

Parasite: a plant or animal that lives off a living "host" without contributing anything to the host's survival.

Pesticide: a substance used to kill animals considered to be pests, such as weevils.

Photosynthesis: the process by which plants use energy from the sun to make their food from water and carbon dioxide.

Pollen: a fine, powderlike substance produced by flowering plants that contains male reproductive cells.

Polyp: a small, round sea animal without a backbone that lives with others of its kind in a colony; the outer skeletons formed by coral polyps gradually form coral reefs.

Quarantine: a period of time in which something is isolated to check for any contagious diseases or other problems.

Recycling: using something more than once.

Savannah: an area of mostly flat grassland, often dotted with trees or shrubs, found in tropical or subtropical regions.

Sealant: something that acts to keep a closure airtight.

Sensor: a device that responds to a signal or some other stimulation, like a rise in temperature.

Septic: relating to the decay of once-living material.

Silicone: a compound, either natural or man-made, with a wide range of uses, including binding other materials together.

Soil-bed reactors: a box of soil through which air is pumped in order to clean that air.

Spaceframe: a very strong steel tube that is the basic building block of Biosphere 2.

Species: a group of closely related living things that are able to breed with each other.

Specimen: a sample used to represent an entire set of something.

Threshing: the process of separating seeds or kernels from stems or husks.

Tissue culture: the growth of tissue cells in a bath of nutrients.

Toxin: a poisonous substance.

Tropics: the area of Earth centered around the equator, usually hot and often with heavy rainfall.

Vertebrates: animals with backbones.

Index

(Page number in *italics* refer to relevant illustrations and captions.)

acid rain 9, 82
Adey, Walter 49
Africa *9*, 47, 48, 54, 67
agave *40*, 45, *54*
agriculture 21, 61
  See also farm
air pressure 22, *76*
Alaska 72
algae 16, 31, *52*, 67, 75
algae-scrubber system 52,
  52, 75
Allen, John *26*, 27, 28
alligator *37*, 48
Alling, Abigail *26*, 27, 28,
  49, *68*, 69, 71–72, *72*
Alvarez, Norberto *26*, *79*
Amazon *44*, 45
ammonia *14–15*
amphibians 41, 42, *46*, 47
  See also frogs, geckoes,
  toads
analytical laboratory 59,
  76, 78
anemones 50
Antarctica *9*, 13, 70, *71*
antelope 37, 38, 48
ants 11, *13*, 42, *42*
architects *22*, 57
architecture 21
apartments *56*, 57–58
Asia *36*
atmosphere 11, 13, 15, 16,
  31, 32, 78
Augustine, Margret *22*, *83*
Australia 48, 60, 69
azolla 67, *67*

bacteria 11, 52, 58
bananas 47, *61*, 62
barrel cactus 53, *54*
bathrooms, *See* waste
  matter
bats 40, *40*, 41, 47
  leaf-nosed bats 41

beans 62, 67
bears 37, 72
bees 26, 40, 42, *42*
Belgium 69, 74
Bios-3 18, *18*
black-necked garter snake,
  See snakes
blue chromos, *See* fish
blue-tongued skink 41
boojum 53
brazil nut 37, 38
bromeliads 46
buffalo 48
Burgess, Tony 53, *54*, 55
bushbabies, *See* galagos

carbon dioxide 14, *14–15*,
  15, 28, 31, 34, 35, *44*,
  76, 78
Caribbean Sea *48*, 50, *50*
carnivores *17*
ceiba trees 46
Central America *36*, 74
cheetahs 47
chickens 10, *64*, 65, 67
Chile 53
China 67, *67*
cleaner wrasse, *See* fish
cloud forest 46, *46*
clownfish, *See* fish
cockroaches 42
  See also insects
Command Center 59, *59*
compost 73, *74*
computers 29, *32*, *69*, 70,
  77, 78, *79*
convection 15, 29
cooling system 35
  See also convection
coral 20, 49, 50, *51*
  coral polyps 50, *51*, 52
crabs *48*, 49, 74
crop rotation 61, *61*

deer
  mouse 37–38
  white-tailed *36*
desert 10, *10*, 21–22, 25,
  29, *30*, 31, 34, *34*, 35,
  35, *40*, 43, 45, 47, 52,
  53, 54, 55, *55*, 57, 63
detritivores *17*
diversity of species 21
dolphins 52
drinking water 59, 63

earthworms, *See* worms
ecotones 55
electricity 10, 29, *29*
elephants 47
endangered species 13
  extinctions 21
energy system 29, *29*
England *22*, 60, 69, *72*
Environmental Research
  Laboratory 61
estuary 48, 52
exercise 59

farm 10, *10*, *20*, 21, 31,
  35, 49, 57, 58, 59,
  60–64, *64*, 65, *66*, 67,
  67, 73
fertilizer 60, 67
finches 40, 48
fish
  blue chromos *50*
  cleaner wrasse 50
  clownfish 50
  four-eye butterfly fish
    20, *50*
  French angelfish *20*
  porkfish *50*
  tilapia, *see* tilapia
Florida Everglades 48
Folsome, Clair 16, *16*, 19
food 18, 49, *56*, 60, 65–67
  growing of 27, 62–63
  preparation of 62, 67

food web 16, *17*, 37
France 60
frogs 46, 47, 49
   canyon treefrog 41
   white treefrog *46*
fruit 17, *20*, 47, *61*,
   62, 63

galagos (bushbabies) 41,
   42, *43*, 47, 55
gallery forest 55
geckoes 46
generators 29
   wave generators 52
Germany 69
giant clams 77
ginger·shrubs 47
giraffes 7, 47
glazing 22, *23*
global warming 9, 82
goats 10, *20*, 65, *65*,
   66, *66*
grasses 47
Great Flood 7, 9, 39

halophytes 49
harpacticoid *31*
Hawes, Phil *22*, *83*
Heraclitus 70, 71, 74
herbivores *17*
Hodges, Carl 61
Hodges, Roy Malone *76*
horses 37
hulling 73
Human Habitat 10, *24*, 56,
   57-60
humidity 25, 27, 54, 60, *76*
hummingbird 7, 38-40,
   38, *40*
   legend of 39

Indian Ocean 71
Insectary 24
insects *13*, *17*, 32, 40, 42,
   42, 47, 49, 70, 74
   See also ants, bees,
   cockroaches, ladybugs,
   termites, pests
Intensive Agriculture
   Biome, *See* farm
irrigation 59, *62*

jojoba 45

karoo rose *20*
keystone predators 74

ladybugs 42, 63, *63*
latex 45
Leigh, Linda 28, *28*, 45,
   65, *68*, 72, *72*, 77
library *56*, 60
life-support systems 13,
   18, 69, 72, *72*, 77
lions 47
lungs 10, 22, *23*, 24

macaque *36*
   See also monkey
macaws *36*
MacCallum, Taber 68, 69,
   69, 70, 72, *72*, 75, 76
Madagascar 55
mangrove trees *48*, 49, *49*
marigold plants 63
marine biology 71
Mars *11*, 18, 82, *83*, 84
marsh 10, 21, 27, 32, 35,
   48-49, *48*, *49*, 52, 72, 76
   mini-marsh 58, *58*
mechanical system 29, 74
medical
   medical clinic 59
   medical procedures *18*,
   27, *28*
   medicines 43
Mexico 50, *51*, 53, 54,
   55, 69
   Baja California *53*, 54
   Gulf of Mexico 50
microbes 11, 12, 16, *17*,
   31-32, *31*, 82
Mission Control *24*, 77
monitoring system *76*,
   77-78, *79*, 81
   sensors *76*, 77
monkey 38
   See also macaque
mosquitoes 47
mosses *45*, 46
moths 40, 41
mouse deer, *See* deer
music 75, 77

National Aeronautics and
   Space Administration
   (NASA) 19, *58*, *81*

Nelson, Mark *68*, 69, 72,
   72
nerve system, *See*
   monitoring system
nitrogen *14-15*

ocean 9, 10, *10*, 13, 16,
   21, 32, 49-52, *51*, *52*,
   70, 71, 74, 75, 78
Ocean, Instant 52
omnivores *17*
orchids 46
Orinoco *44*
overpopulation 9
oxygen 13, 14, *14-15*, 28,
   34, 58, 76, 78
oysters *48*, 49
ozone depletion 82

Pacific Ocean 52
palm trees 13
   peach palms 46
panama hat plant 45, 46
papaya *20*, 46
peanuts 61, 62
pests 63, *63*
   control of 60
photosynthesis 14
pigs 10, 62, 65, 66, *66*, 67
Pima Indians 39
pollination 38, *38*, 39-41,
   40, 42
   pollen 38, *38*
pollution 9
polyps, *See* coral
potatoes 27, 62
Poynter, Jane 60, *63*, 66,
   68, 69, *73*, 77
Prance, Ghillean *22*, 38
Puerto Rico 60

rainforest 10, *10*, *20*, 21-
   22, 29, *30*, 32, 35, 41,
   44, 45-46, 47, *49*, 53, 70
recreation 59, *75*
red-spotted toad, *See*
   toads
reptiles 41, *41*, 42, 47
   See also snakes,
   tortoises
rice *61*, 62, 67, *67*, 73
robot builders 82
rubber tree 45, 46

Santa Catalina Mountains
2, 24
savannah 10, 21, 32, *32*,
35, 41, 47, 48, 52
Schneider, Silke *65*
sealant 22, 26-27
sea urchins 52
seaweed 52
sensors, *See* monitoring
system
septic system 58
See also wastewater
sharks 52
sheep 7, 67
silicone 22
Silverstone, Sally *59*, *68*,
69, 72, 73, *73*
snails 47
snakes 41, *41*, 47
black-necked garter 41, *41*
soil-bed reactors 31, 63
solar power 10, 29
Sonoran Desert 2, 30, 57
(Arizona desert), 72
South America *36*, *44*, *47*,
48, 53
southern spadefoot toad,
See toads
Soviet Union 18, 19
space 82
space explorers *12*
spaceframe 22, *23*, 27
space shuttle *81*
space stations 13, 63, 80
space travelers 81

Spider Woman 39
starfish 52
Storm, Stephen *12*
sugar cane 63

television 75
temperature 15, *21*, 29,
29, *30*, 35, 54, 60, 76,
76, *79*
See also air pressure
termites 11, 26-27, 42, *42*,
47, 48
Test Module *24*, 25, *25*,
26, 27-28, *28*
thornscrub forest 55, *55*
threshing 73
tiger *37*
tilapia 67, *67*
tissue culture 12, *12*, 63,
64, *64*, 81, 82
toads
red-spotted 41
southern spadefoot 41
tortoises
yellow-footed 41, *55*
transition zones 55
See also ecotones
tree ferns 46
tropics 29
tropical garden 27
tropical orchard *20*
tropical plants 43, *45*
turtles 41
sea turtle *50*
See also tortoises

United States 24, 60, 69, *72*

Van Thillo, Mark *68*, 69,
70, 72, 73-74, *73*, *75*
Venezuela 32, *45*
Venus *11*
Vernadsky, Vladimir *19*
vertebrates 41
Vertebrates X, Y, and Z
27-28, *26*, *28*
video 60, 75

Walford, Roy *60*, *68*, 69,
70, *72*, *73*, 75
waste matter *17*, *18*, 27
wastewater 27, 58-59, *58*
See also septic system
wave generators, See
generators
weather 32, 34
whales *37*, 71, 82
wheat 62
whirlygig beetles 42
worms 11, *17*
earthworms 47

yellow-footed tortoise, *See*
tortoises
Yugoslavia 69

Zabel, Bernd *67*
zebras 47